Lamborghini: Cars And Tractors.

**Ferruccio Lamborghini has been a tractor manufacturer and had a great dream: creating the perfect car.
He did it in Sant'Agata Bolognese, close to Modena...do you remember Ferrari?**

Any color but red: that is Lamborghini's choice. Even in its choice of colors, Lamborghini's innovative nature is evident.

It was the first automaker not to espouse a hue for the sake of recognizability, but equally it knew how to be exceptionally recognizable.

The more colorful sports cars were, the more recognizable they were as Lamborghini.

The colors, all of them, were, Lamborghini!

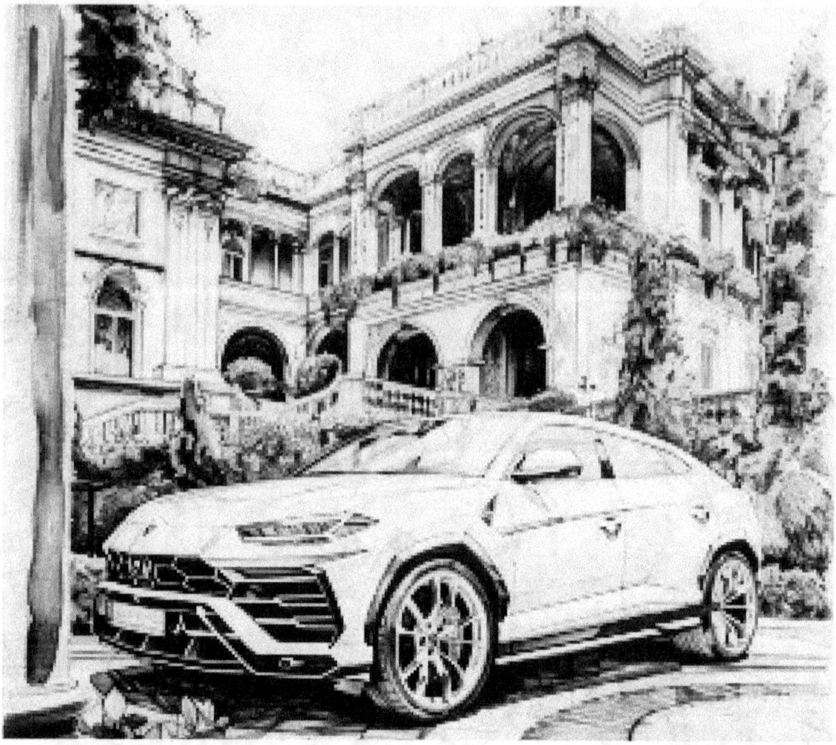

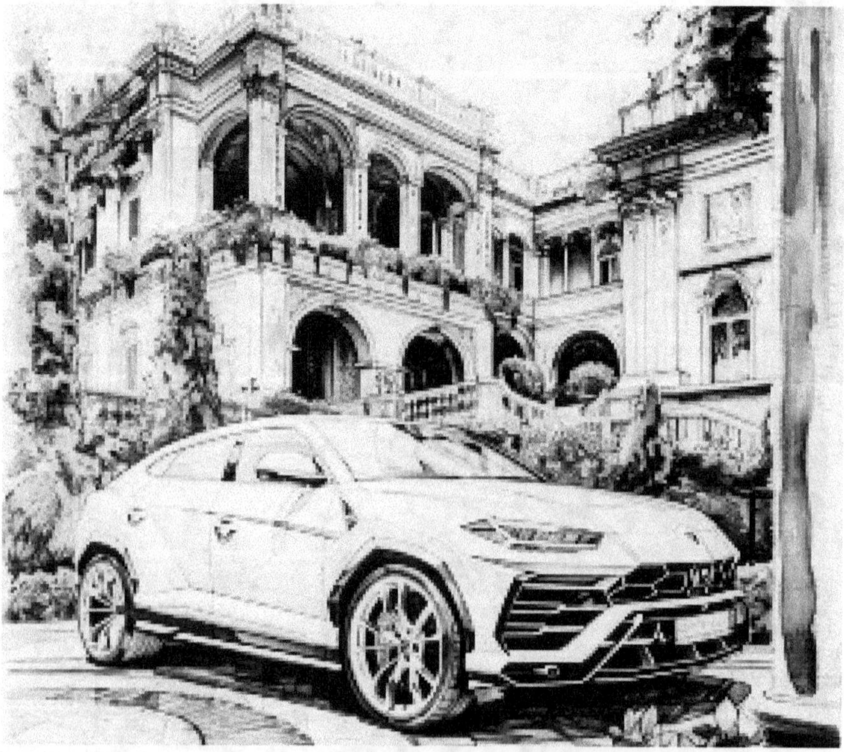

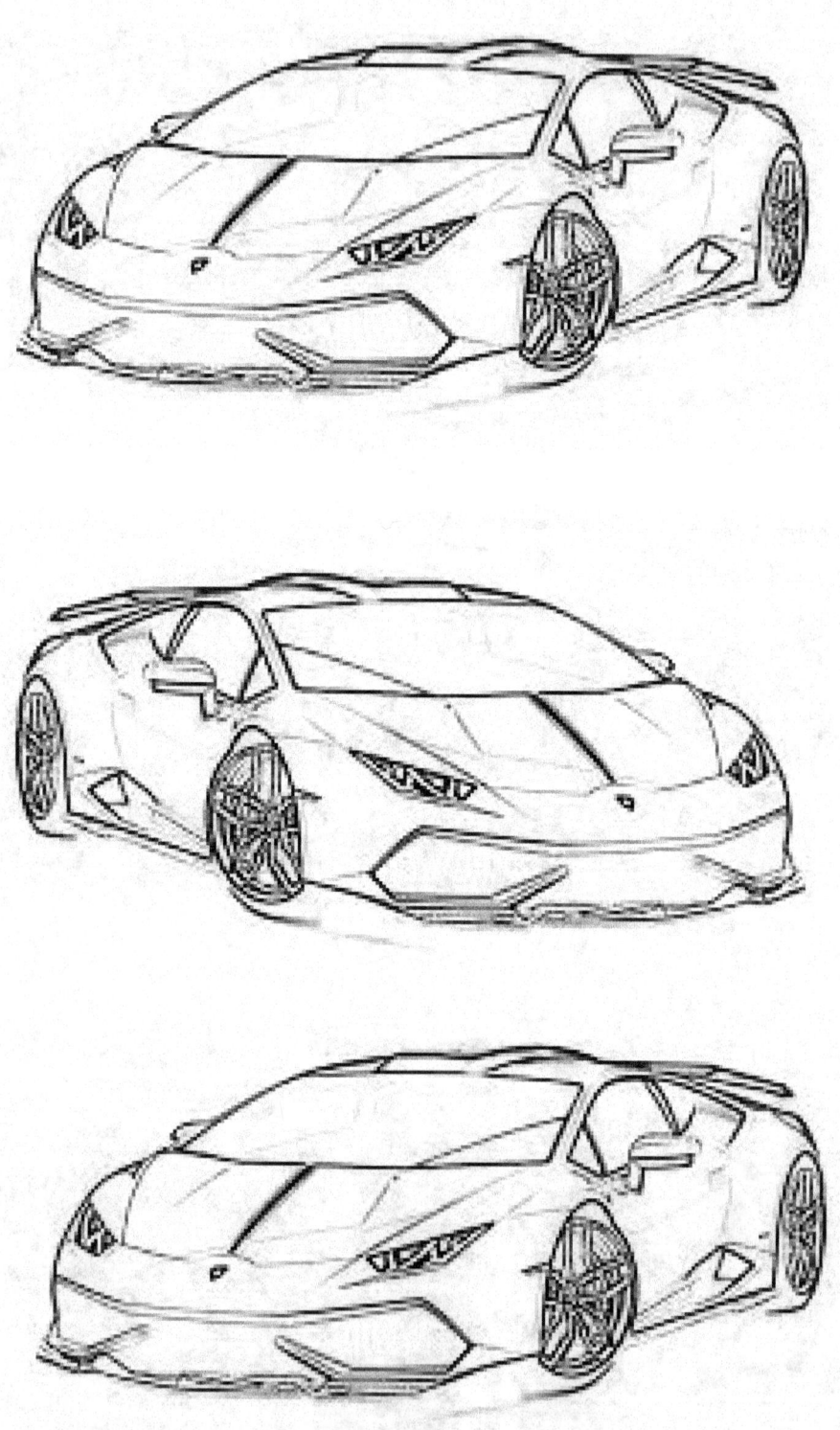

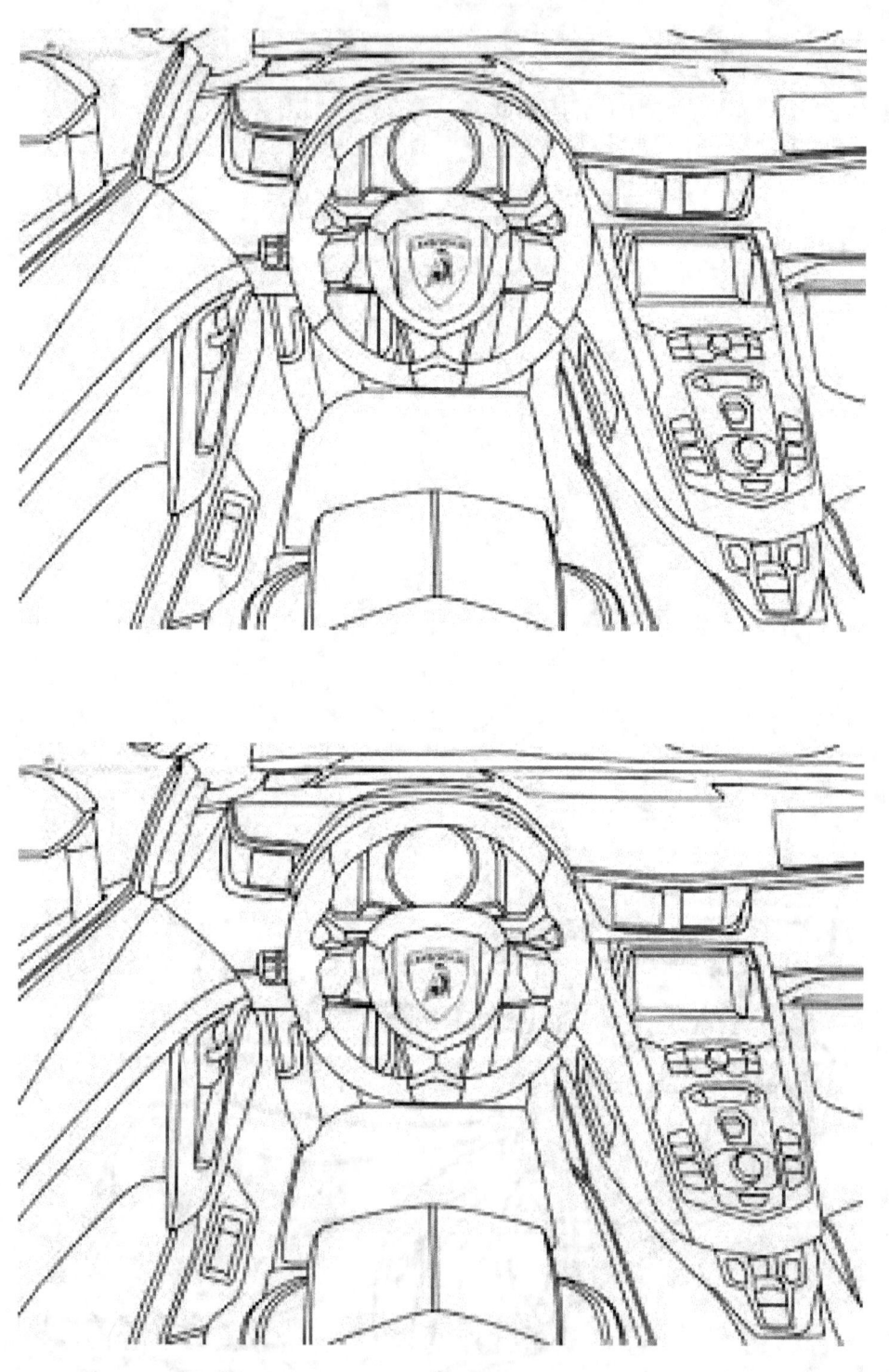

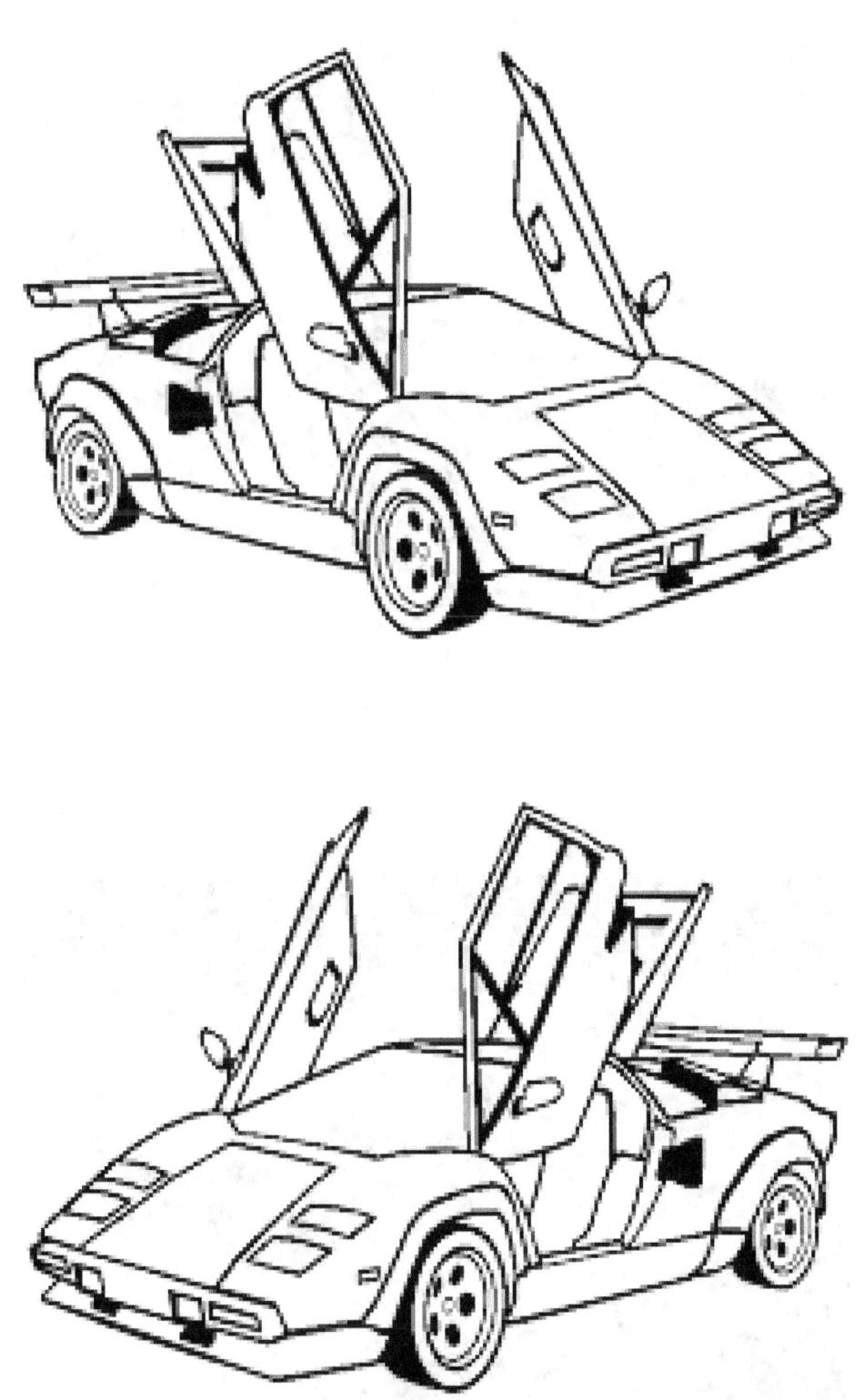

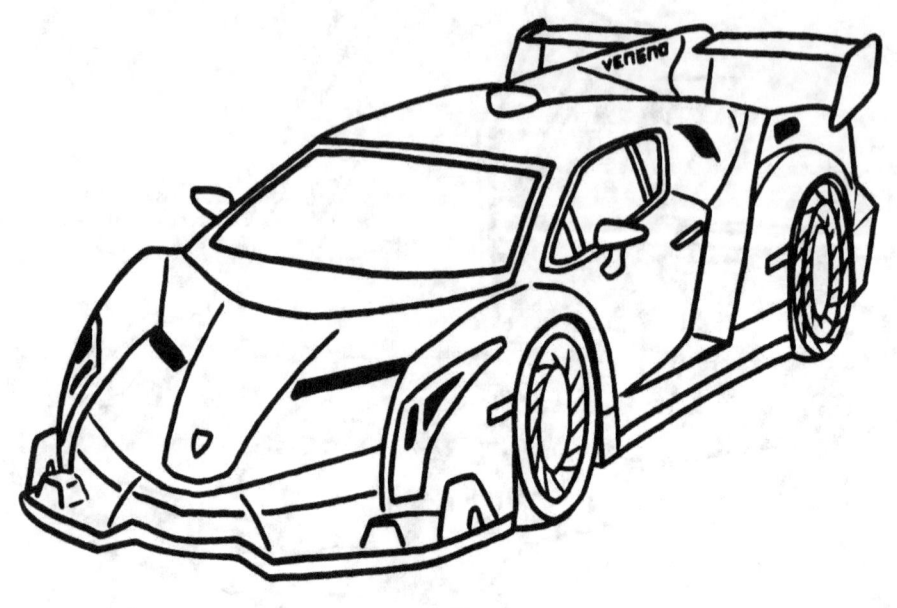

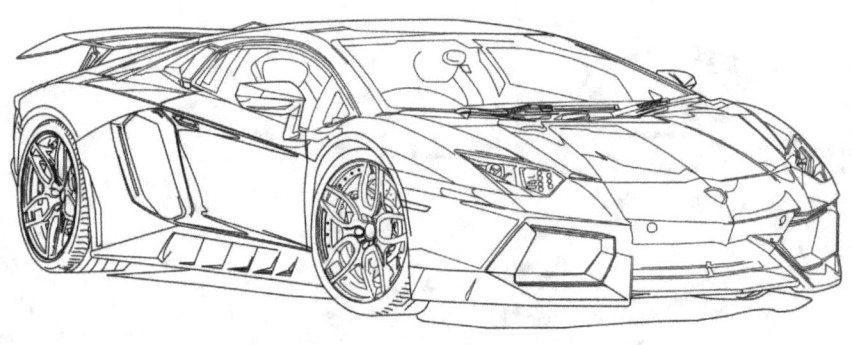
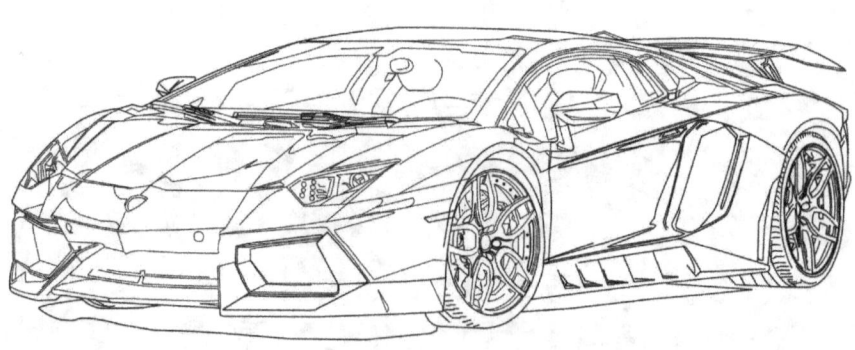

I really do not know if the choice of the symbol was a message of defiance to Ferrari. Taurus was the zodiac sign of Ferruccio, born on April 28, 1916.

The bull is a stronger animal than a horse...Lamborghini was always in competition with Ferrari anyway.

Consider that he chose his suppliers by following a single, unbreakable rule: no one who worked for Ferrari.

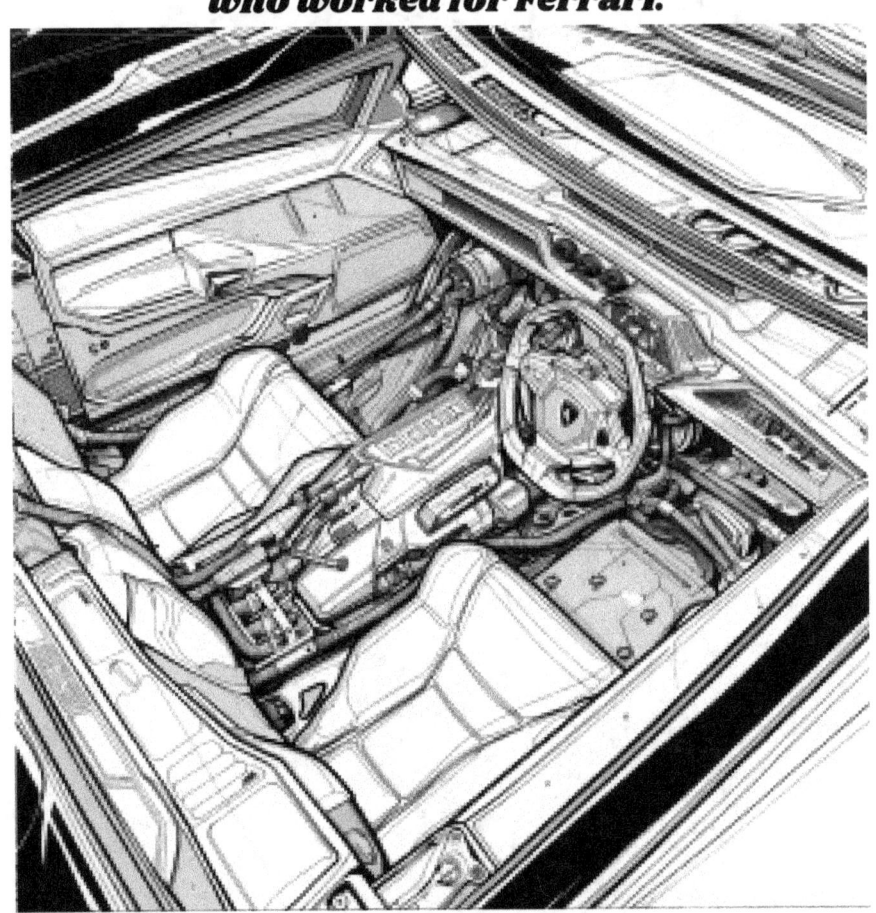

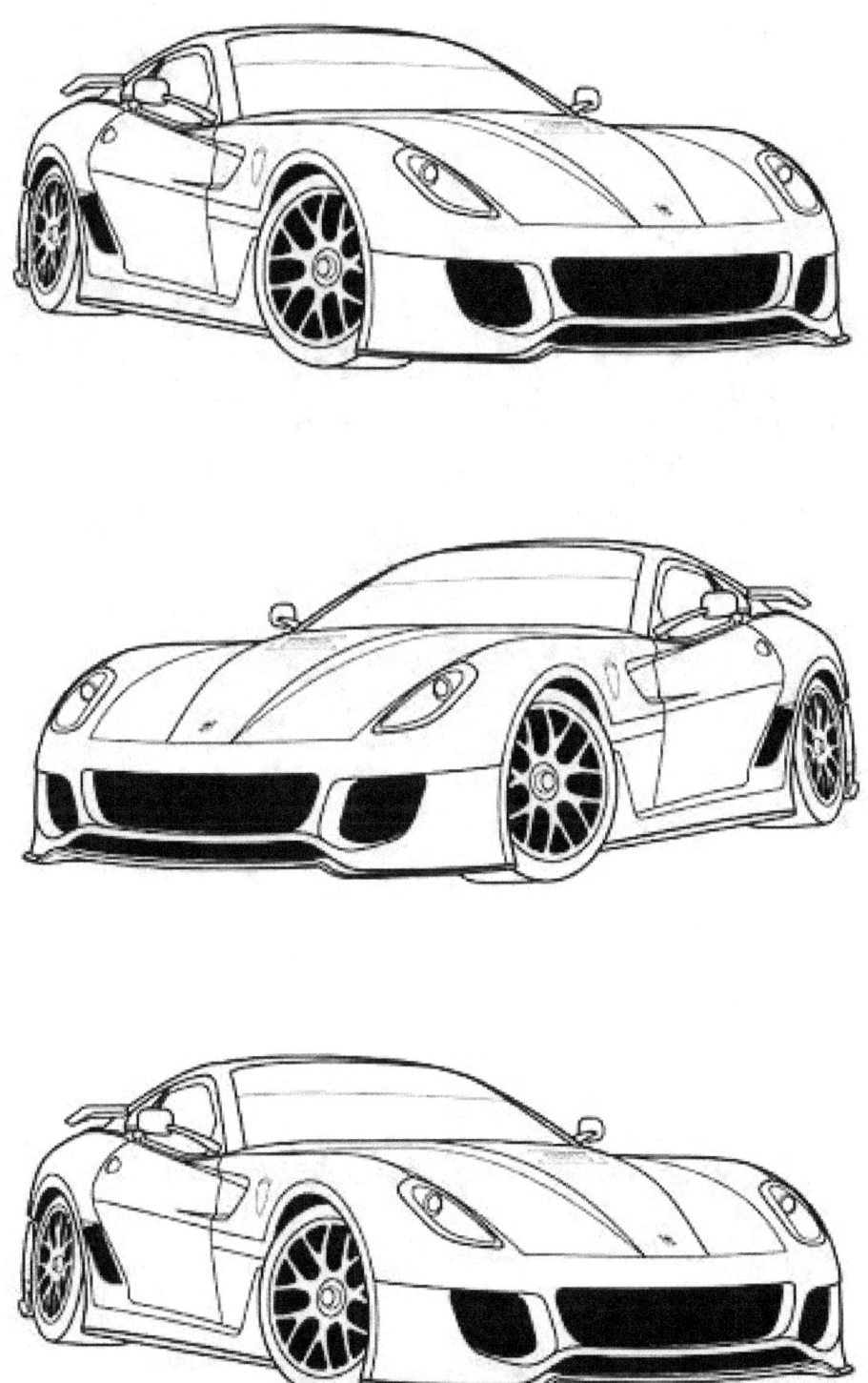

In Rome...

Lamborghini: "I would never want a driver to lose his life in one of my cars."
On the one hand the desire to enter formula 1, on the other the desire to make his cars, the cars for happiness!

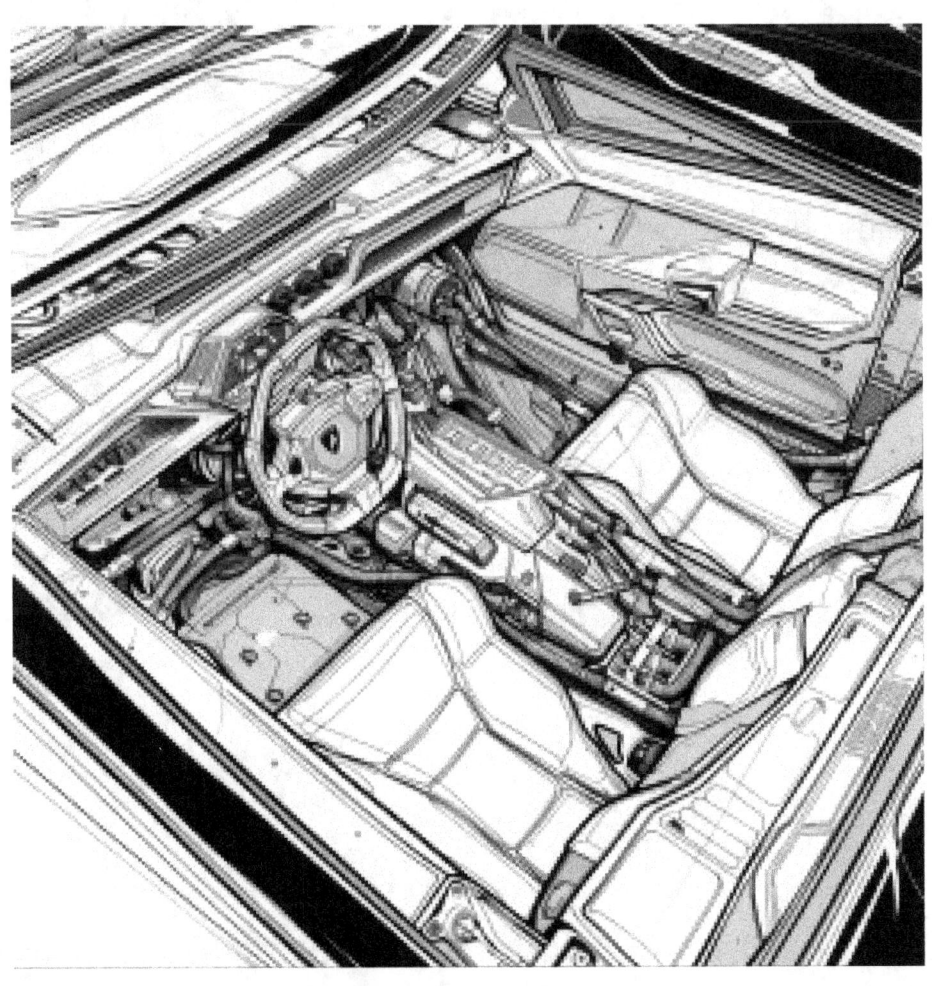

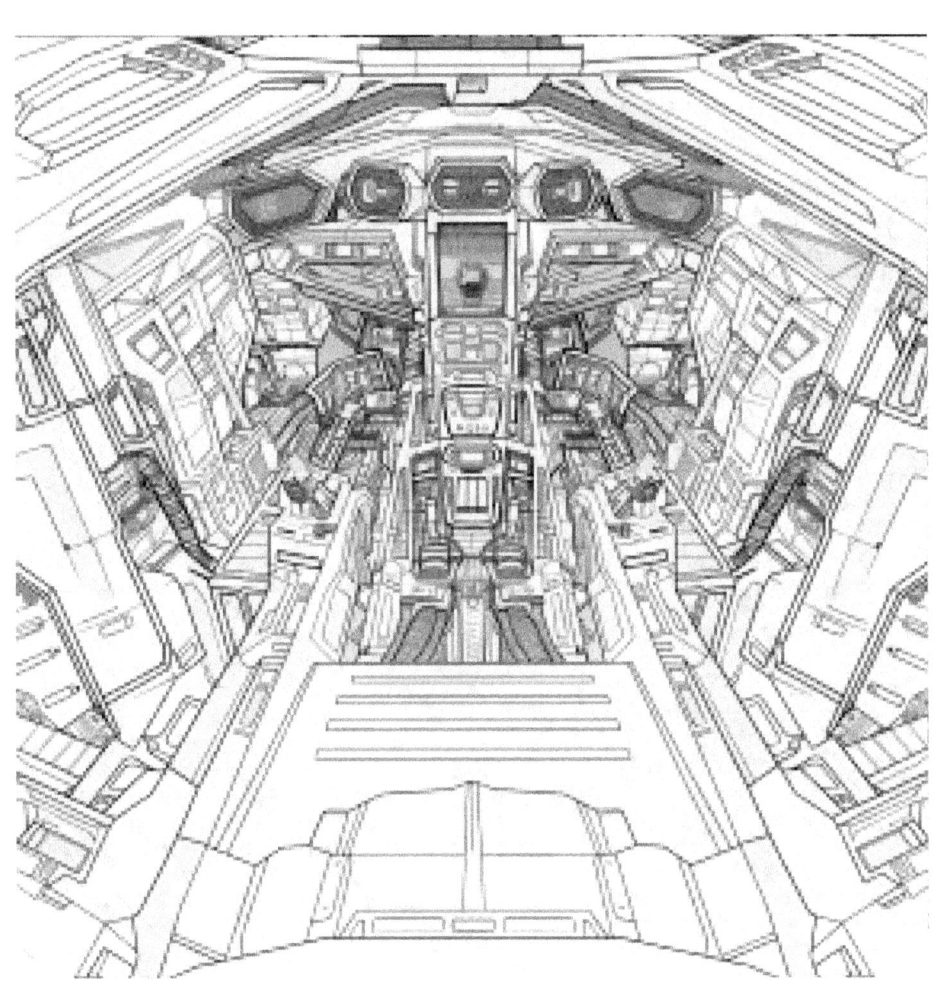

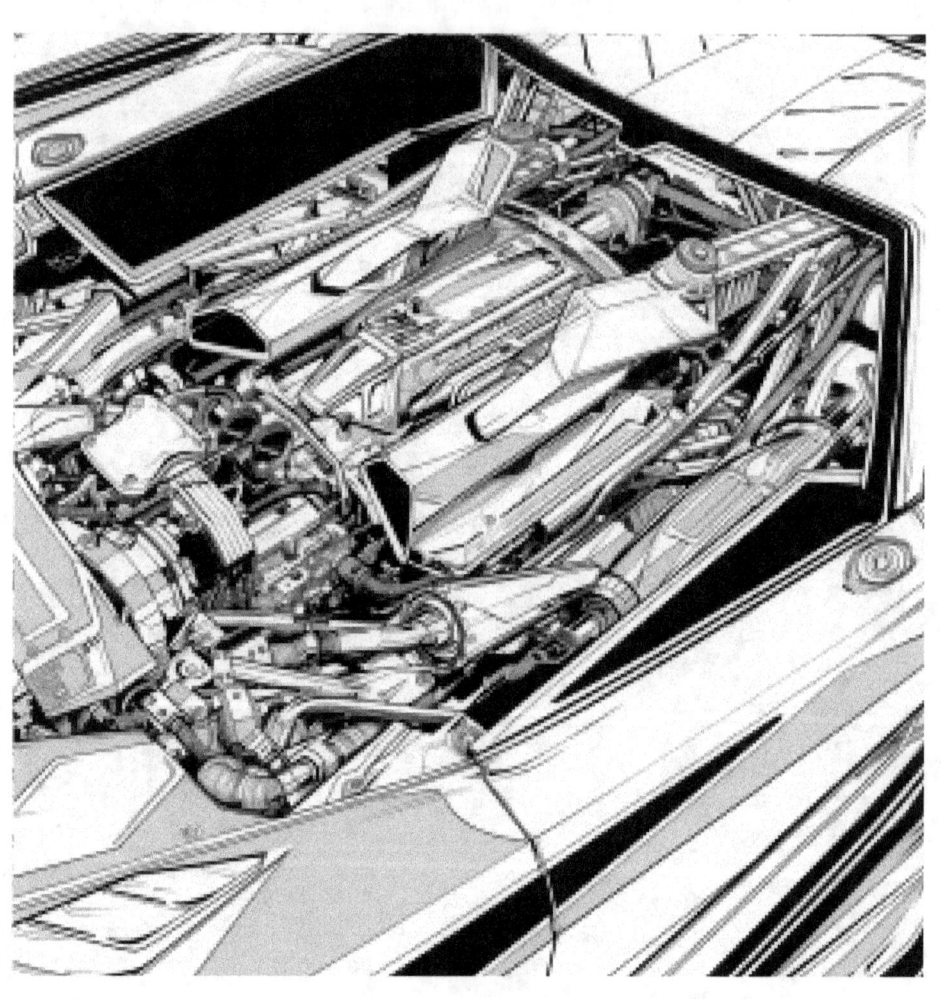

In Pisa...

In 2018 Lamborghini creates the Urus and doubles its small female customer base, which is only 11 percent of the total.

In 1986 Lamborghini produces the LM002, an off-road vehicle designed for desert territories.

300 examples are sold and are now cult objects among a very particular kind of costumers (from Stallone to Saddam Hussein)

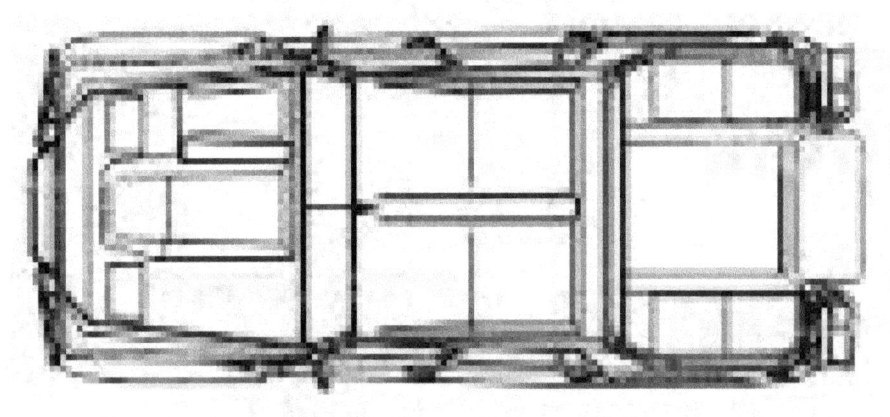

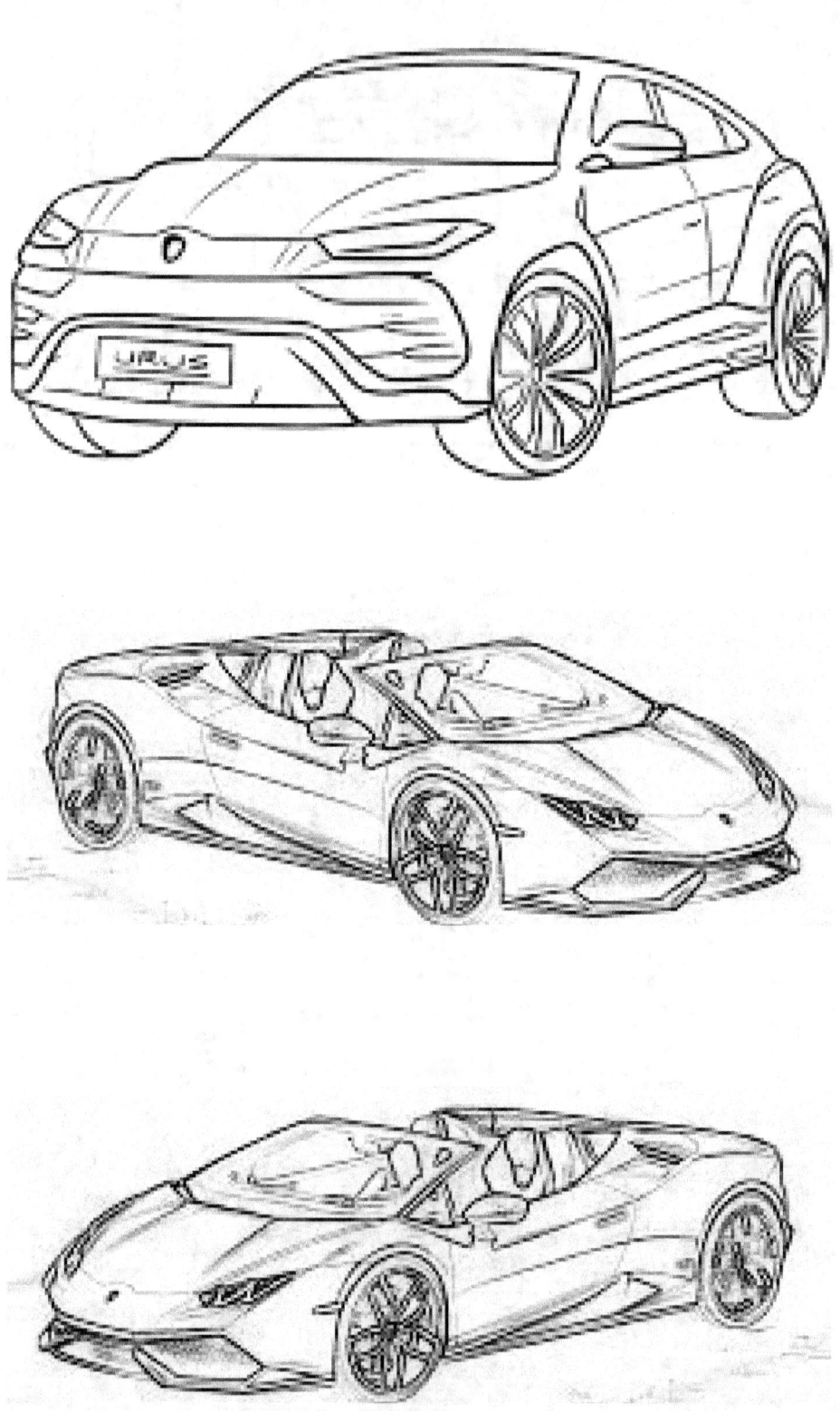

The only example of the Miura Roadster made in 1968 is owned by New York collector Adam Gordon,

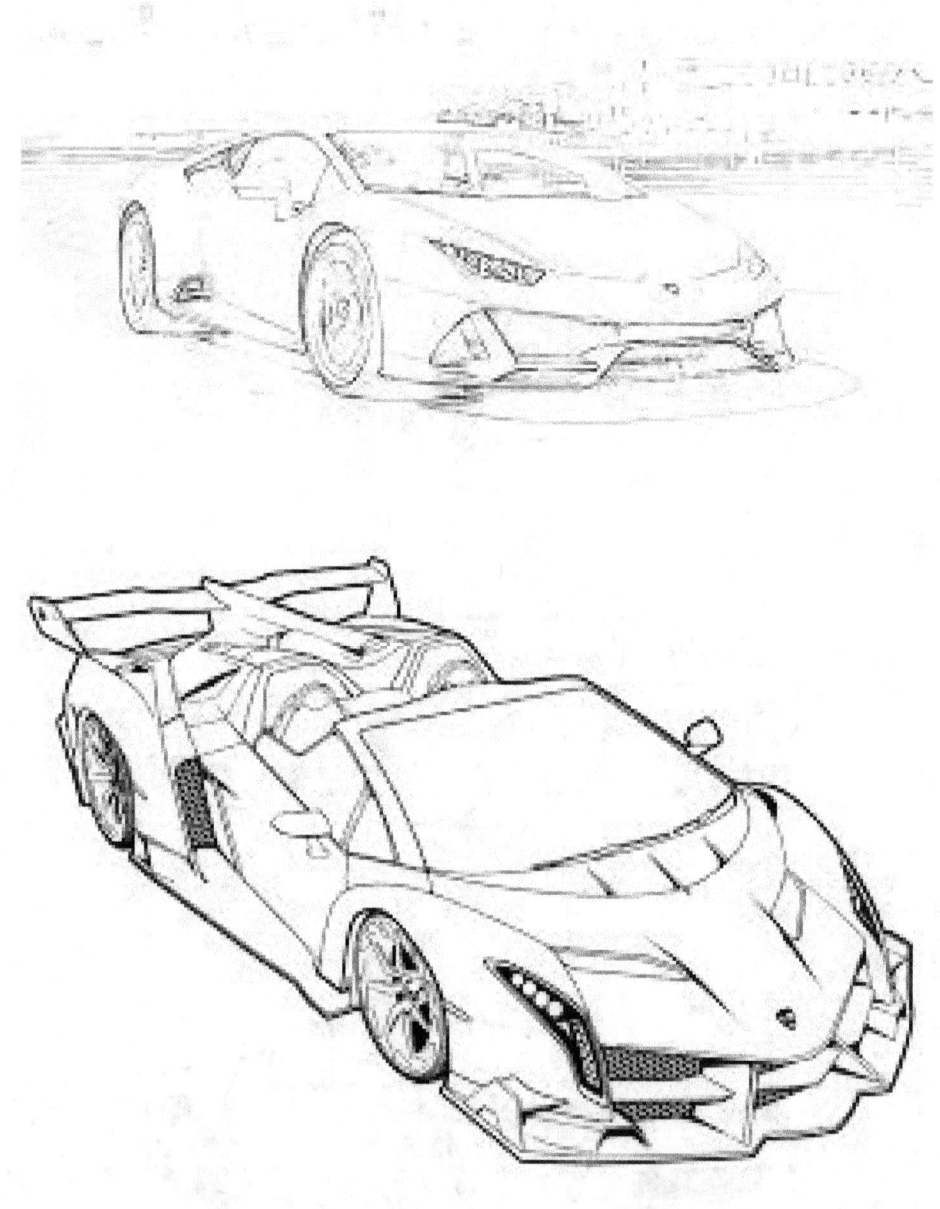

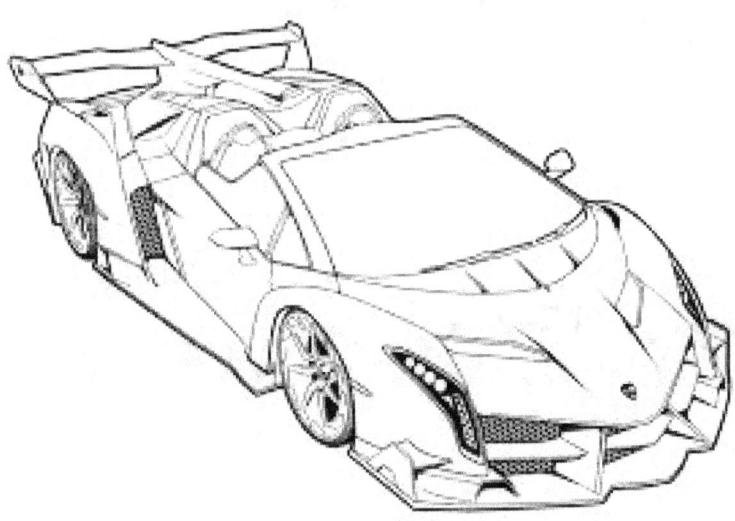

In Florence...

Lamborghinis have spanish names that refer to bullfighting.

A great fan of bullfighting, Lamborghini took inspiration from bovine nomenclature.

Strength, power, and speed are the hallmarks of these beautiful machines.

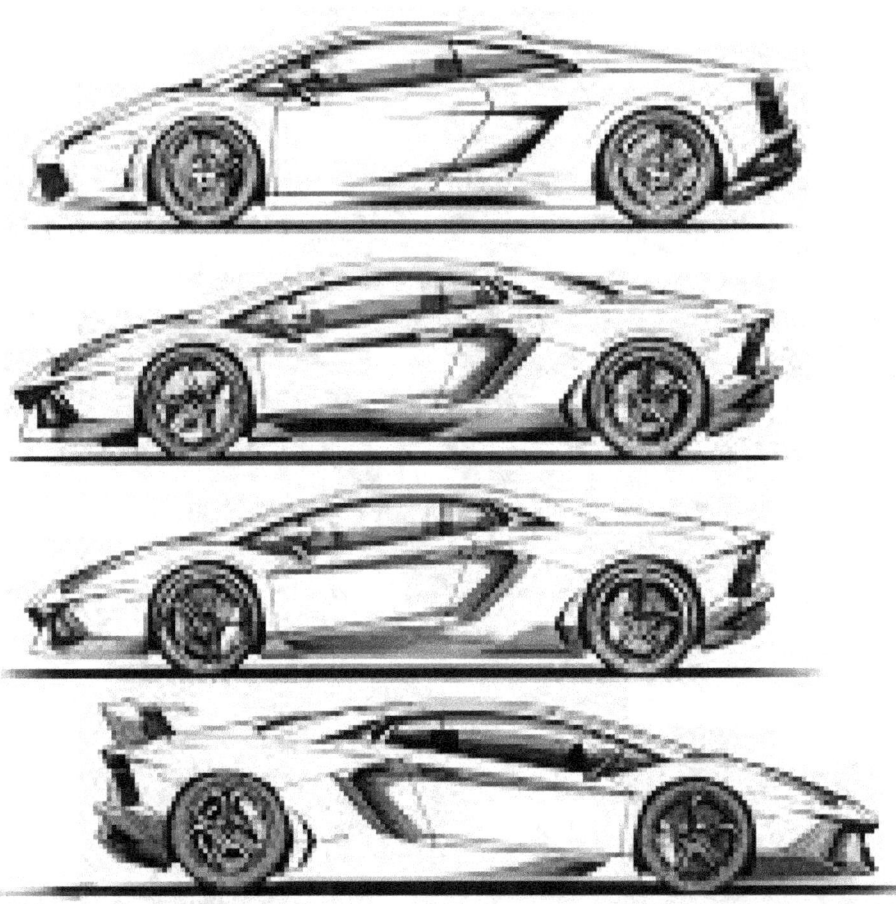

The valuation record at an auction for a Lamborghini was hit in September 2019 in Geneva....8 million.

The Lamborghini Veneno, of which only nine examples of the roadster version were produced, was made in one year, between 2013 for the company's 50th anniversary celebrations.
The previous record was for the sale of a 1972 Miura SV sold in Paris for $2.5 million.

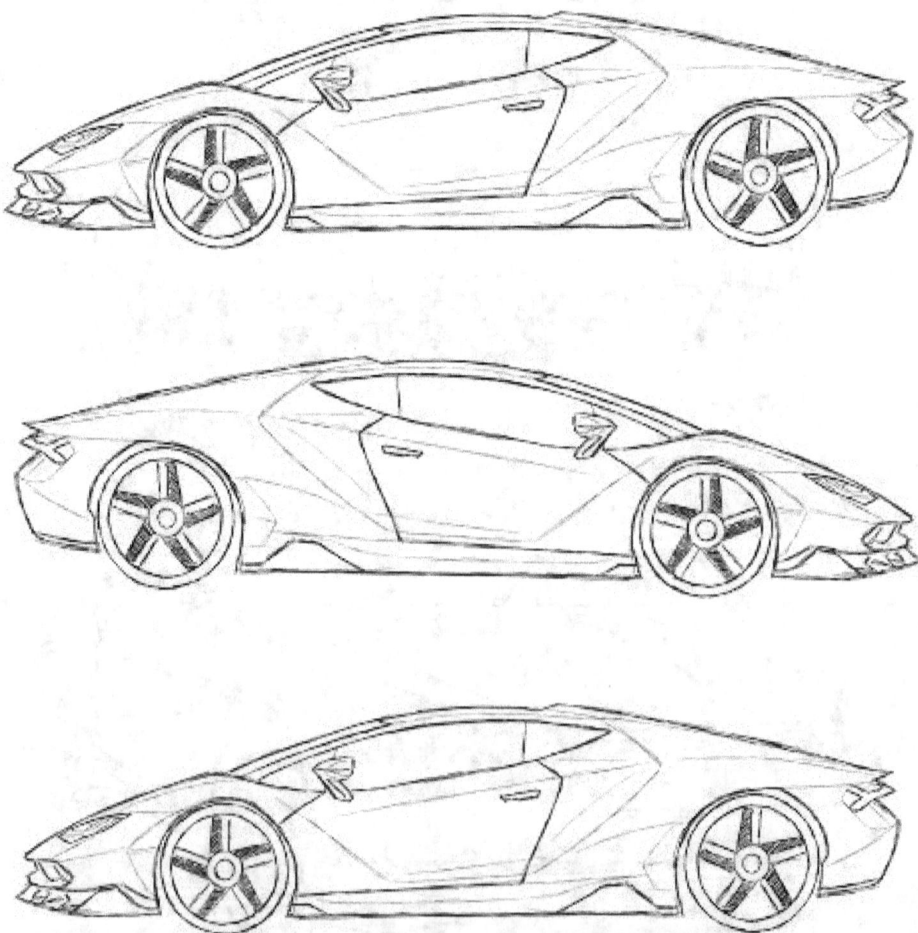

Lamborghini, an iconic Italian sports car manufacturer, was founded by Ferruccio Lamborghini in 1963.

The company was established with the aim of creating cars that could rival Ferrari's, sparking a legendary rivalry.

Lamborghini's first model, the 350 GT, was introduced in 1964, showcasing the brand's commitment to design, luxury, and performance.

Over the years, Lamborghini has produced some of the most iconic sports cars, including the Miura, Countach, and Diablo, each pushing the boundaries of automotive engineering and design.

The brand is known for its distinctive styling, often characterized by sharp lines and aggressive forms, as well as its use of powerful V12 engines.

Lamborghini has also made significant contributions to motorsports, competing in various racing series and developing high-performance engines.

Today, Lamborghini continues to innovate, blending traditional craftsmanship with cutting-edge technology to create some of the most coveted supercars in the world.

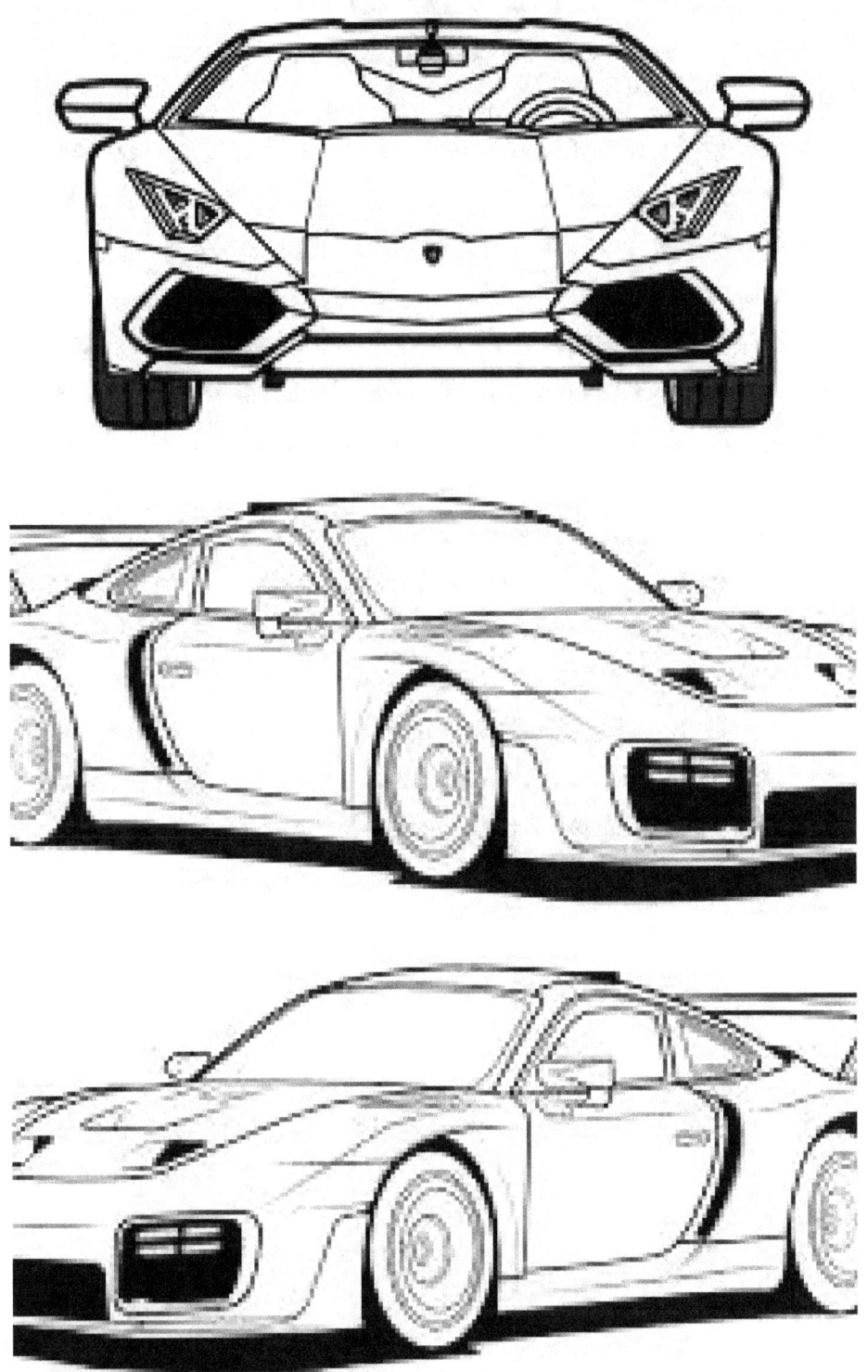

Ferruccio Lamborghini wanted to create the perfect sports car - it was his dream!

The story goes that the brand's founder offered engineer and designer Giotto Bizzarrini a bonus every time he managed to add power to the new 3.5-liter V12 engine, mounted at 60° with twin overhead camshafts and capable of delivering 320 hp on the 350 GT: Lamborghini tradition was born; the dream was realized.

The first model was the basis for further iterations of the V12, with numerous refinements developed in subsequent models: the 400 GT (1966), the Espada (1968), the Islero (1968) and the Jarama (1970), the latter evolving into the 365-hp Jarama S, Ferruccio Lamborghini's favorite.

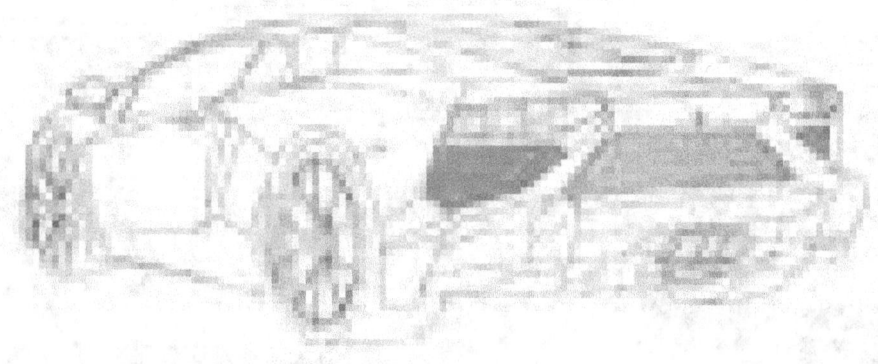

The Miura, born in 1966 and evolved into the 1971 SV, turned all the rules upside down: the V12 was moved to the center-rear of the car, providing better weight distribution and achieving 370 hp, 0-100 km acceleration in 6.7 seconds and a top speed of 285 km/h, and turning the model into the fastest production car in the world at the time of its launch.

The unmistakable sound of the V12 is like a symphony, a sweet sound.
Each of the 12 cylinders must move in tune with the others, like a series of violins playing in crescendo to best enhance the driver's driving sensations.

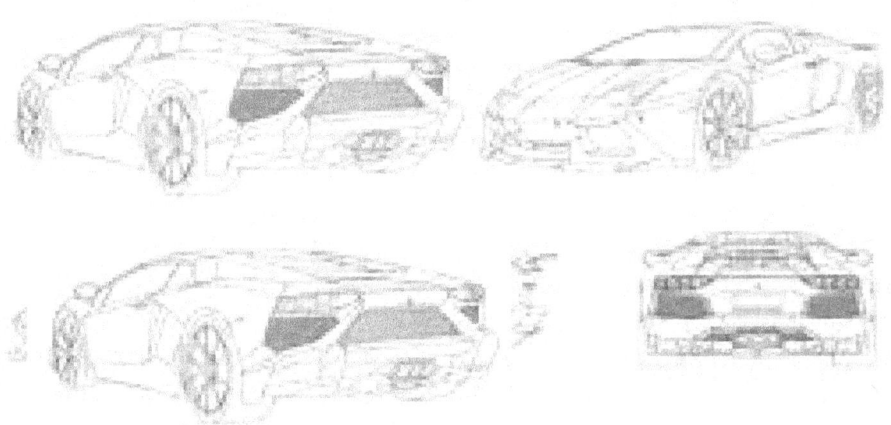

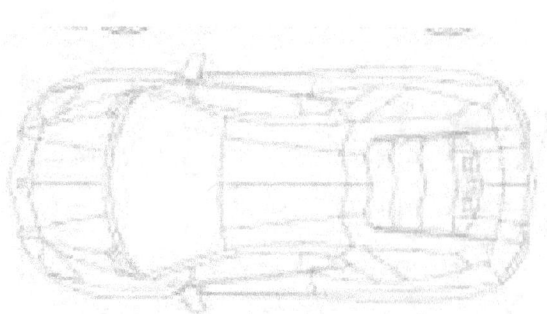

FERRARI'S HISTORY IS A TALE OF PASSION, INNOVATION, AND UNPARALLELED SUCCESS IN THE WORLD OF AUTOMOBILES AND MOTORSPORTS. FOUNDED BY ENZO FERRARI IN 1939 AS SCUDERIA FERRARI, THE COMPANY INITIALLY FOCUSED ON RACING AND SPONSORING DRIVERS BEFORE TRANSITIONING INTO MANUFACTURING ITS OWN CARS. THE FIRST FERRARI-BADGED CAR, THE 125 S, WAS INTRODUCED IN 1947, POWERED BY A V12 ENGINE THAT BECAME A HALLMARK OF THE BRAND. THROUGHOUT THE 1950S AND 1960S, FERRARI ESTABLISHED ITSELF AS A LEADER IN GRAND PRIX RACING, WINNING NUMEROUS CHAMPIONSHIPS AND SOLIDIFYING ITS REPUTATION FOR HIGH-PERFORMANCE ENGINEERING.

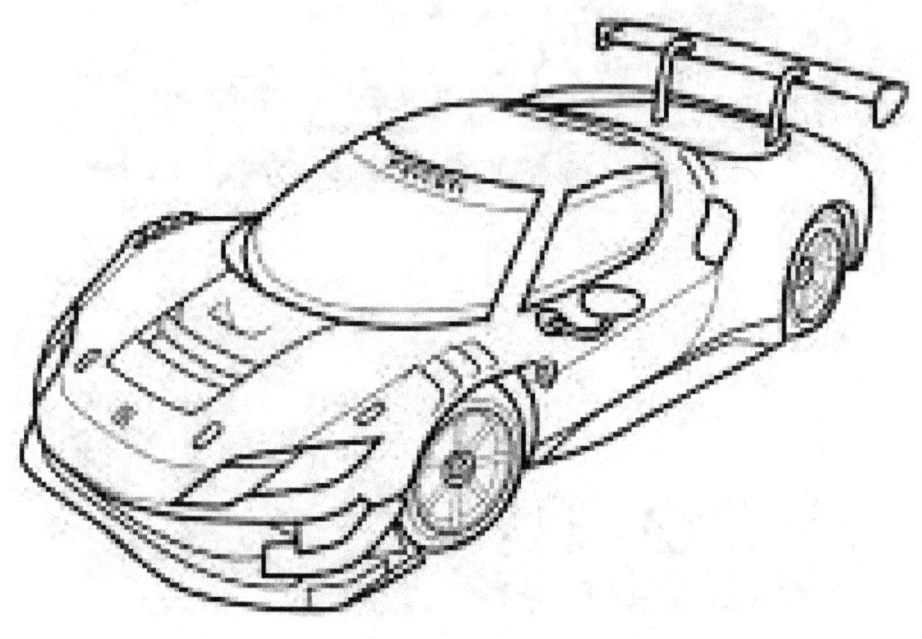

THE BRAND'S COMMITMENT TO INNOVATION WAS EVIDENT IN MODELS LIKE THE 250 GTO, A SPORTS CAR THAT COMBINED RACING PROWESS WITH LUXURY, BECOMING ONE OF THE MOST SOUGHT-AFTER COLLECTOR'S ITEMS. FERRARI'S SUCCESS CONTINUED WITH THE INTRODUCTION OF MODELS LIKE THE DAYTONA AND THE DINO, EACH PUSHING THE BOUNDARIES OF DESIGN AND TECHNOLOGY. THE PRANCING HORSE LOGO, DERIVED FROM THE INSIGNIA OF WORLD WAR I FLYING ACE FRANCESCO BARACCA, BECAME A SYMBOL OF SPEED AND EXCELLENCE. TODAY, FERRARI REMAINS AT THE PINNACLE OF THE AUTOMOTIVE WORLD, BLENDING TRADITION WITH CUTTING-EDGE TECHNOLOGY TO CREATE SOME OF THE MOST COVETED SPORTS CARS, WHILE MAINTAINING ITS LEGACY AS A DOMINANT FORCE IN MOTORSPORTS.

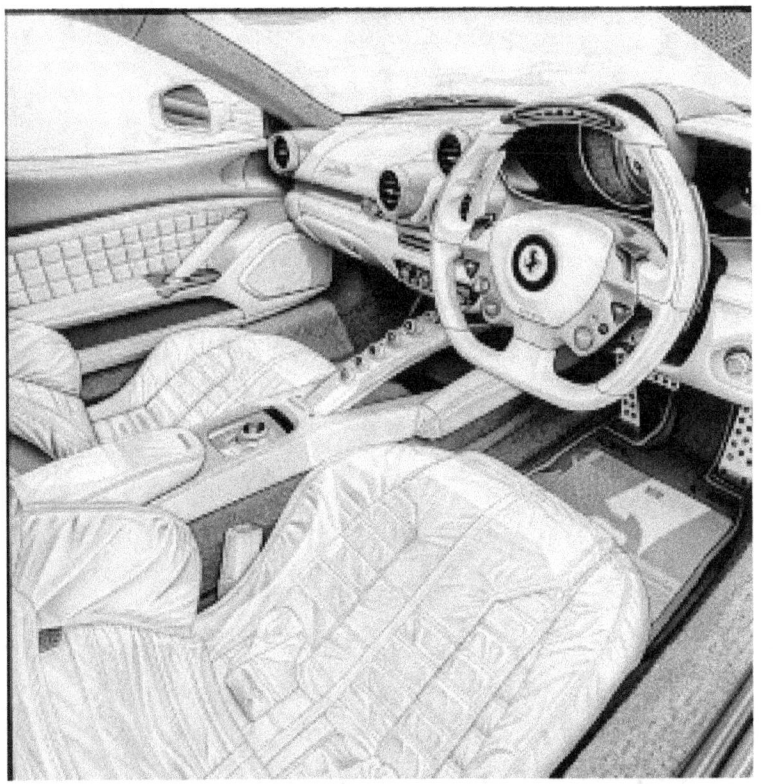

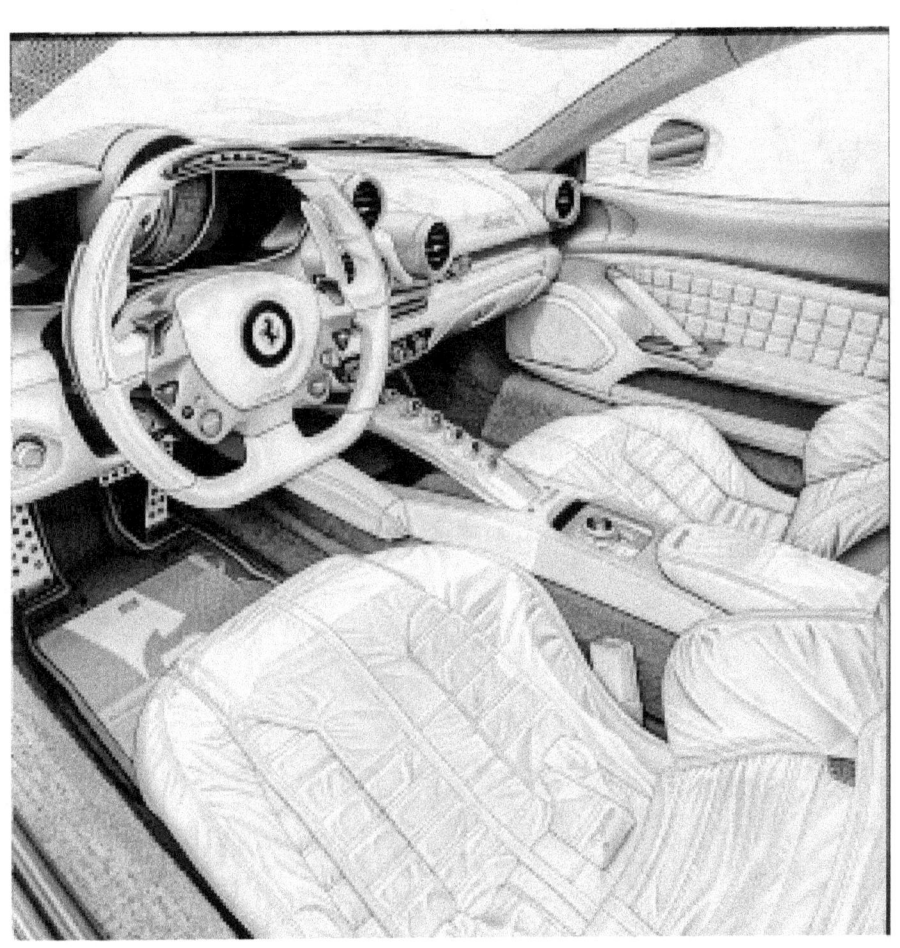

Arrivederci!

Goodbye!

www.ingramcontent.com/pod-product-compliance
Lightning Source LLC
Chambersburg PA
CBHW052339220526
45472CB00001B/488